Albert Reid Ledoux

The Sugar Beet in North Carolina

Report to the Commissioner of Agriculture on the Results of Experiments

with the Sugar Beet

Albert Reid Ledoux

The Sugar Beet in North Carolina
Report to the Commissioner of Agriculture on the Results of Experiments with the Sugar Beet

ISBN/EAN: 9783337144395

Printed in Europe, USA, Canada, Australia, Japan

Cover: Foto ©berggeist007 / pixelio.de

More available books at **www.hansebooks.com**

THE SUGAR BEET

IN

NORTH CAROLINA.

———•———

REPORT TO THE

COMMISSIONER OF AGRICULTURE

ON THE RESULTS OF

Experiments with the Sugar Beet

IN THE COUNTIES OF

ANSON, BEAUFORT, BURKE, CABARRUS, CHATHAM, DUPLIN,
EDGECOMBE, GRANVILLE, ORANGE AND WAKE,

BY

ALBERT R. LEDOUX, Ph. D.,

CHEMIST TO THE DEPARTMENT OF AGRICULTURE,

AND

Director of the State Experiment and Fertilizer Control
Station, at the State University.

CHAPEL HILL, N. C.

———▬———

RALEIGH:
FARMER AND MECHANIC STEAM BOOK, JOB OFFICE & BOOK BINDERY.
1878.

REPORT.

Hon. L. L. Polk, *Commissioner of Agriculture,*

Dear Sir: I have the honor to lay before you the following report upon the Sugar Beets grown in this State, and sent to the Experiment Station for analysis. It is a matter of profound regret to myself, as it is to you and the other members of the Board of Agriculture, that so few and meagre returns have been made for the trouble you have taken, and the seeds sent out.

The correspondence of those experimenters who have made any report will point out many of the causes of failure. Chief among them, perhaps, is the fact that the seeds did not reach their destination until late in April, or early in May, a full month too late. There are other causes undoubtedly which had their influence, and prominent among them, no doubt, was a lack of *full* knowledge as to the proper means of cultivation. With this in view, I have embodied in my report at some length, a synopsis of the best German experience. Owing to a desire to awaken an increased and intelligent interest in the subject, I have also added a brief resume of the present state of the Sugar Beet industry in this country, and some other matters which may be found practically valuable.

Though the results of your efforts to have the State fairly represented have been so unfortunately few in number, yet the analyses of the product of the ten counties represented in my report are by no means discouraging. While in no case does the per centage of sugar reach 12, yet the average of all is much higher than that obtained in some States, and quite encouraging. At any rate I hope that the present report may be the means of keeping up the interest in the State, and that by beginning in season, and with the help of this year's experience, and with new light on the subject, we can show the world next year that the results we hoped to attain can be fully realized.   Yours respectfully,

ALBERT R. LEDOUX,
Agricultural Experiment Station,
State University, Chapel Hill, N. C.

January 9th, 1878.

SUGAR.

Sugar, in the form with which we are most familiar—the so-called "Cane Sugar"—has been known and used from the most remote ages in India and China, the very name coming down to us through the Arabic or Persian language, and it is known as "Sukkar" in Arabia at the present day. The "Calamus," and "Sweet reed" of the Bible are also supposed to refer to the Sugar Cane.

The manufacture of Sugar came slowly into Europe, entering by way of Venice in the 10th century. Strabo, Arrian, Pliny and others had already mentioned in their historical accounts of the nearer Orient, the occurrance of a plant—undoubtedly the Cane—which yielded a syrup that was eaten as honey with bread, and was brought originally from India and Ethiopia.

Pliny says further that it was called "Sacoharum," and that sometimes when allowed to flow from the bruised plant it would form a white, solid substance resembling salt, which was used as a medicine. The early crusaders found the Syrians indulging in a sweet juice "extracted from a Cane which they broke up in mortars, and sometimes allowed this extract to stand in the sun and evaporate, when a whitish substance separated out, which was eaten with bread." The crusaders got some of the seed, and bringing back samples of the Cane, they introduced its cultivation into Rhodes, Sicily and Crete in the 9th century. Thus spreading from the Levant as a starting point, the process of manufacture reached Venice in 996, Spain and Portugal coming next, and finally in 1319 Sugar became an article of importation into Great Britain in considerable quantity.

It is by no means improbable that the Spaniards found the Sugar Cane already growing, when they discovered the West Indies; at any rate with their wonderful adaptability of soil and climate, and the subsequent introduction of slave labor, they soon came into complete control of the sugar markets, and in the 16th century India, Europe, and the Mediterranean Islands were driven out of all competition, and their manufactures languished.

There are three chief saccharine substances, differing slightly in chemical composition, which are more or less familiar to us. These are called "Cane Sugar," "Grape Sugar," and "Milk Sugar." The last gives to milk its sweet taste, and is found only in that animal secretion of which it constitutes from 3 to 10 per cent. It is made from whey, on quite a large scale among the mountain dairies of Switzerland, and finds its chief use as a vehicle for Homeopathic medicines, and in some localities as an article of food. It is white, hard, and brittle.

Grape Sugar, called also "Glucose," is undoubtedly the most abundant and widely distributed in nature of the three forms of sugar. It gives to almost all fruits their sweet taste, and is the main cause of the sweetness in nearly all our cultivated vegetables. It can moreover be made artificially from starch by a very simple process, and yields readily to fermentation, forming Alcohol, and on this account it is coming more and more into demand for the manufacture of beers and alcoholic liquors. It is not crystalizable.

Cane Sugar is to every one a familiar friend, and needs no description. It is the most common of all our so-called luxuries; the last we give up when compelled to economize. *It is claimed by some political economists that the con-

---

* In 1866, at the close of the war, the consumption of sugar per capita in the United States was only one half what it was in 1876.

sumption of Sugar will give a very fair idea of the wealth and prosperity of a people.

Unlike Grape Sugar, Cane Sugar is produced by comparatively few plants, in sufficient quantity to render its extraction profitable. The Sugar Cane, Chinese Cane, (or Sorghum) the Sugar Maple, a few species of Palm and the Sugar Beet being the only members of the vegetable kingdom from which it is obtained in any quantity. Nor can it be made artificially. Of the above mentioned sources of Cane Sugar, Sugar Cane supplies 66 per cent., Sugar Beets 28 per cent., the Palms 5 per cent., and the Maple 1 per cent.

In the Report of the United States Commissioner of Agriculture for 1876, we find the following tables, showing the consumption, source and cost of the sugar used in this country.

SOURCE AND CONSUMPTION.

" The commercial estimate of the supply of the past year is as follows :

	TONS.
Cane Sugar, domestic and foreign	638.869
Cane Sugar received on the Pacific coast	28,800
Cane Sugar made from Molasses	43,600
Maple Sugar	13,000
Domestic Beet, Sorghum, etc.,	2,000
Taken for consumption in 1876	725,269
Taken for consumption in 1875	773,002

On the basis of a population of 45,000,000 the consumption would be 36 pounds to each in 1876, and 38 for the population in 1875. The sugar supply of the commercial world in 1875 was 3,457,623 tons, of which 40 per cent. was Beet Sugar made in Europe. Cuba produced one-third of the Cane Sugar; the other West India Islands and Brazil,

Java and Mauritius, are all prominent sources of supply. The [following is an estimate from high authority of the quantities produced of both kinds in 1875:

CANE SUGAR.

	TONS.
Cuba	700,000
Porto Rico	80,000
British, Dutch, and Danish West Indies	· 250,000
Java	200,000
Brazil	170,000
Manila	130,000
China	120,000
Mauritius	100,000
Martinique and Guadaloupe	100,000
Louisianna	75,000
Peru	50,000
Egypt	40,000
Central America and Mexico	40,000
Reunion	30,000
British India and Penang	30,000
Honolulu	10,000
Natal	10,000
Australia	51,000
Total tons	2,140,000

BEET-ROOT SUGAR.

	TONS.
German Empire	346,646
France	462,259
Russia and Poland	245,000
Austria and Hungary	153,922
Belgium	79,796
Holland, and other countries	30,000
Total tons	1,317,623

COST.

" The cost of these sweets is a serious burden upon the country. ' We have the soil to produce a full supply, either of cane or beet sugar and laborers suffering for work, and measures should be taken for a rapid increase of home production. The details of the cost of the sugar used in this country, subject to a slight reduction from re-exportation, are thus given in the statistics of the customs receipts:

FISCAL YEAR OF 1876.

Sugar, brown	pounds,	1,414,254,663	$55,702,903
Sugar, refined	pounds,	19,031	1,685
Molasses	gallons,	39,026,200	8,157,470
Melada, syrup, &c	pounds,	79,702,878	2,415,995
Candy, &c	pounds,	87,955	18,500
			$66,296,553'

Beside the cane proper, and sugar beet, a few other sources of sugar have been suggested or tried in this country. Among the most prominent being sorghum, the maple, and recently watermelons.

SORGHUM.

Quoting again from the report of the Agricultural Bureau for 1876:

" As an estimate for twenty-one years since the introduction of sorghum, 11,000,000 gallons of syrup per annum might approximate the product. At an average value of 65 cents (it is less now) the value of the annual product would be $7,150,000. The sugar of sorghum is a small item, yet in fourteen years, in Ohio alone, it amounts to

50,000 lbs. Including sugar and forage, the annual value must be not less than $3,000,000, and the aggregate value $168,000,000 since its introduction by the Department of Agriculture."

MAPLE SUGAR.

This industry which is of less importance in our Southern country, is nevertheless of considerable value to the United States. The total amount of sugar and syrup obtained from this source in 1870 was equivalent to about 57,000,000 pounds of sugar, which at 10 cents per pound gives a total value of $5,700,000.

The manufacture of sugar from watermelons is of more interest to our people than that from the sugar maple, and we shall watch with interest an experiment now in progress in California. A stock company, with a capital of $2,000,-000 is about commencing operations, and though chemists and manufacturers are rather doubtful of their financial success, they enthusiastically claim that they can obtain 10 per cent of sugar from the juice, alcohol from the pulp and rind, and 25 per cent. of oil for table use from the seeds.*

THE SUGAR BEET.

Having thus briefly examined the other sources of cane sugar, let us now turn to the Sugar-Beet.

As long ago as 1747 a German chemist discovered the presence of cane sugar in the white and red beet, and in 1796 the first factory for the manufacture of beet sugar was established in Prussia. The great cost of cane sugar made the new idea of obtaining it from a domestic source exceed-

---

* At some future day I hope to be able to make some experiments with the Sweet Potato, which has a large per centage of grape sugar and starch, and may yet possibly form the basis of a large industry in North Carolina and the other Southern States.

ingly alluring. The experiment spread through Europe, with many a failure, many a lesson gained through the loss of enormous fortunes invested; now taxed, now protected by the Governments, with constant improvements in cultivation and machinery, until at the present day nearly one-third of the cane sugar used by the civilized world is obtained from beets. At first the percentage of crystalizable sugar in the juice of the beets was low, and only with improved means of cultivation, the results of many experiments, did the French and German agriculturists learn to produce a uniform average of 12 per cent. or over.

The beet is a natural growth in several localities, abounding in a wild state on the Mediterranean coast. The present varieties of sugar beet are the result of cultivation and hybridization.

Before speaking of the results of our experiments in this country to raise the beet profitably, I have deemed it best to present to our farmers a synopsis of the results obtained by the European experimenters, and which show what treatment the beet requires from the cultivator to give uniformly good results. This information is clearly and concisely stated in Dr. Stammer's "Lehrbuch der Zuckerindustrie," and I beg to be allowed to give a somewhat free translation of the valuable chapter on the

CULTIVATION OF THE BEET.

1. " *The Soil.*—Although neither by chemical analysis nor by examination of the physical properties, can we tell in every case that a certain soil will or will not grow the beet successfully, yet experience has shown that, in general, successful culture requires a soil *loose; deep, rather more rich in humus; more loamy and limey than sandy ; with porous subsoil, and a warm, sunny exposure. Of course not deficient* in any of the necessary ingredients of plant food, which may

easily be the case, when the potash and phosphoric acid have been too largely drawn upon.

The recognition of a suitable soil for sugar beets presents greater difficulties than for many other plants, for they obtain most of their indispensable nutriment, by means of their long root, from the sub-soil, and the composition of this sub-soil is therefore of immense importance. It is on this account that the experiments hitherto, with superficial manuring, have yielded no uniform results. We manure that portion of the soil, to be sure, from which the growing beet does long derive nourishment, but not that portion whence the plant obtains its food during the all important period of the formation of sugar. And the chemical means of sending the manures down into the sub-soil (viz: by admixture of chloride of sodium,) are by no means so certain in their application that we can trust confidently that invariable results will follow every such experiment.

On the other hand, and in perfect accord with this, *deep plowing*, (subsoiling) has given the best and surest results in beet culture ; and *all* observations upon the happy influence of the steam plow upon the beet crop, without exception, (if we look at them in the proper light) may be referred back to this cause, *deep plowing.*

From this standpoint *all those efforts which have for their aim the improving of the subsoil by mechanical, as well as by chemical means are the most important in beet cultivation.* In other words, on one hand the deeper cultivation of the ground, on the other the sub-soiling ['unter grunddungung,' (manuring the subsoil.)]

Chemical analysis of that portion of the soil which we are accustomed to call the sub-soil ('acker krum') with a view to the cultivation of sugar beets, save in exceptional cases, is of little or no importance or use. And as far as the physical properties are concerned, *experiment is always the best means of ascertaining whether a soil is suited for beet culture or not.* Of course such soils as do not meet the general requirements mentioned above are out of the question; for example, such as are too sandy, wet or stony. And on the other hand, those soils which, from their origin would be expected to possess those elements of plant food most abundant in the ash of the beet, will more probably show a better adaptability for beet culture. We should not, however, draw too hasty conclusions from the result of a single experiment. The work expended upon the soil becomes perceptible only by degrees, hence a field only becomes a good beet-growing field by degrees."

MANURING.

2. " Manuring should always first of all give back to the ground what the harvest has removed from it, and not only the mineral (inorganic) constituents, but also the nitrogen. *Nothing is surer than that a soil to which a full return of plant food is not made, loses by degrees its power to produce the crop required in normal quantity and composition.* The experimental cultivation of the beet with artificially prepared fertilizing liquids has been much less pursued than with other plants, and therefore the relation between the composition of these liquids and the development of the beet is not yet determined. We lack also the basis upon which to predicate the direct action of manures upon the beet. Here lies the difficulty, above indicated, of applying the manurial substances to that layer of the soil whence the beet principally derives its nourishment. Hence in the present state of our knowledge and of our fertilizers, the object of our fertilization can be nothing more than the *retaining in good condition of a soil which is already suitable for beet culture.''*

" After the above remarks it will not be thought astonishing when we say that all the laborious and painstaking experiments with the manuring and culture of beets have as yet given no results uniform and everywhere applicable. Such results we can only expect from a study of those laws which may be deduced from the artificial cultivation of the beet in special liquids.

And yet it is in no contradiction of these facts when we advise the beet culturist to keep up constant experiments with fertilizers upon different soils. It is in such cases only necessary to determine the particular form and quantity of manure which under the peculiar local conditions gives the best returns. And in many cases some particular form of manure will prove the best, but the power to produce a safe and invariable influence upon the crop will only seldom be attained. The effect of those factors over which we have no power, climate and weather, are of infinitely greater influence than the small alterations which we can produce by the augmentation, deterioration, or maintenance of good condition of soil within the circumscribed limits of artificial fertilization."

" Experience has taught that those beets which are raised upon fields manured with fresh—especially stable manure are less suited for manufacturing purposes. On this account the rule has long been established that the manure should not be applied directly to the beets, but to some other previous crop, or, that beets should be cultivated as the 2nd or 3rd in a series of rotation. Unfortunately this rule, most important to the *manufacturer*, was not so generally observed in earlier times as it should have been, so that very often on account of heavy manuring large crops were obtained, but *at the expense of the sugar*, or quality of the juice.

This rule is especially applicable to stable manure, and that from cess-pools ; less so to the so-called " artificial fertilizers" which, when they are not employed directly in too great quantities, are followed by fewer injurious effects.

The principal constituents which must be taken into account in reckoning the addition to and removal of plant food from the soil by beets, are Potash, Phosphoric Acid, Mag. nesia and Nitrogen. As the amount harvested differs with the soil and other circumstances, we will therefore employ for our calculations following, the mean proportion of these four substances present in 1000 lbs of beets and beet tops, as determined by analysis :

	1,000 lbs of	
	Roots	Leaves
	CONTAIN :	
Potash	3.9 lbs.	6.5 lbs.
Phosphoric Acid	0.8 "	1.3 "
Magnesia	0.5 "	2.7 "
Nitrogen	1.6 "	3.0 "
Ash	7.1 "	18.1 "

We see from this table, by noticing the proportion between roots and leaves, that the removal from the soil by the leaves is so considerable, that it should receive quite especial consideration in the calculation, when the tops are not returned to the field immediately after the harvest. The latter proceeding is to be urged all the more, since on most beet farms there is a deficiency of fodder, and it is a temptation to replace the loss in fodder by feeding the tops. From this standpoint, the wide spread custom of paying for pulling the beets by giving the tops to the laborer for his work, is an evil which should be striven against. *It is pretty certain that a full compensation to fields so treated cannot be effected.*

"The removal of potash can be easily reckoned, as in the following illustration, for example, and thereby we can show what return of potash is needed, if the field is to continue to produce plants containing potash. In one distillery in France, which is, to be sure, rather exceptional in the enormous business it does, over 82,000 lbs. of molasses per day are converted into alcohol, equivalent to the yearly harvest of 79,000 acres of beets. The residue from this molasses is worked up into potash and soda salts. These salts were originally extracted from the soil in minute quantities, little by little, by the long and tedious processes of vegetation; processes artificially inimitable. They are exclusively used in chemical industries, and not returned to the soil.

If we calculate the amount of potash which is removed from 79,000 acres in the molasses and add to it besides that removed with the raw sugar, we find it reaches at least 28,000 cwt. per year, for which compensation *must* be made.   *   *   *   * As in this case only the potash is considered which was obtained in the *final product,* these figures are much below the reality; really deceptive in fact, when we think of what is lost by imperfect extraction, and left in the press cake, &c., &c.

In this way should every farmer calculate, in order to find out whether there is danger, either in the near or distant future, that his land should become poor in potash. That such a result will happen is certain, even though a particularly bountiful supply of potash in the soil may put it off for some time."

"The *form* in which the above mentioned plant constituents should be returned to the soil, is *fixed* as far as the phosphoric acid and magnesia are concerned; partly also for the nitrogen. Super phosphates, with more or less accompanying nitrogen (naturally present or added) may always be used. The magnesia may come from the waste material of sugar manufacture, with which direct investigation has shown it is nearly all returned to the soil, although the state of sub-division does not insure entirely even distribution. This latter defect may be partly remedied by cutting up or composting. It is to be recommended from time to time to make calculations based on analysis of the manurial substances employed, so as to ascertain the amount of phosphoric acid, and especially magnesia, added to the soil. For these last two substances this (calculation) is easily made. More difficult is the question of the potash which has been removed by the crop. Manuring with potash salts is frequently undervalued, and undoubtedly because large and tangible results were expected which failed to appear, while the *chief end* of potash manures is neither in augmenting nor bettering the crop, but *in causing it to hold its own.* This result is especially noticeable from the fact that no diminution takes place in the yield, which would certainly be the case in a greater or less number of years if the compensation was not complete. * * * * *

A further consideration, and such an one as would greatly modify the results, lies in the *form* of the potash compound employed. There is no other point on which the opinions of practical men so much differ as in this, and continually are new compounds declared to be the best; but of *universal* application alone is the rule above, that we should always mix the potash salts with common salt (NaCl), in order to insure their being conveyed to the lower soil; also the admixture of magnesia salts, when these have not been applied in some other way. None of the potash salts from natural deposits possess any peculiar merit above the others. But those having an admixture of organic matter seem to me to be preferable. * * * * For this purpose, that potash coming from the beet itself—the residue rich in lime, the molasses, &c.—is most valuable and should be returned to the field when possible. One should not believe, however, that potash sufficient for the development of the plant has been added when the molasses and other waste products of the beet harvest have been returned to the field. Without taking into consideration the leaves, which may have been left upon the field, a very large amount of potash is still necessary, and the molasses alone does not restore the amount needed by a good

2

deal, as an easy calculation will show. In manufactories where raw sugar is sold, much potash is disposed of with the sugar, and in all manufactories the waste water always carries off potash compounds, and although in comparatively small amounts, yet in sufficient quantity to account for the difference between the amount of potash found in the beet, and in the molasses. This is no theoretical consideration, but one founded upon exact comparative analyses."

" There are however large tracts of beet growing country where, on account of the present state of things, or owing to their locality, this style of manuring (with beet re-fuse) is difficult or impossible. For such, as also for the ever present deficit above mentioned, we are thrown back upon the "Potash salts," and this *must* be the case on many farms till an easier method of manuring with beet refuse [press cake] is discovered. Without allowing myself to go into the question as to which is the best Potash salt, and why such dissimilar results from manuring are observed, I will nevertheless point out the fact that the universally good results which follow the manuring with beet refuse [press cake] will serve as a kind of guide board for us; that is, that the present method of applying the Potash salts broad cast over the field should be supplanted by another, viz : *dissolving the salt in liquids which are rich in organic matter.*

We should certainly expect that a solution of the Potash salt in the urine from the stalls and stables for example, would insure a very equal distribution of the Potash in the soil, and in fact in a form better suited to the assimilative powers of the plant, than scattering about small crystals of an inorganic Potash compound. Naturally this same result may be reached in other ways, as for example, by mixing a concentrated water solution with the other manure, or with the compost heap and applying to the field the manure thus enriched with Potash. Experience and personal experiment will point out the preferable way.

The advantages of such mixing of Potash salts with the stable liquids (often accomplished by farmers by strewing the Potash salt about the stalls) are thus enumerated by Frank.

1. The sulphate of magnesia contained in the Potash salts holds (retains) the Ammonia and Phosphoric Acid. 2nd. The too rapid fermentation of the urine is prevented. 3rd. The prevention of the loss of Ammonia, and too rapid fermentation make the manure sweeter and more healthy. 4th. The tediousness of scattering broadcast is obviated, and a much better sub-division and distribution upon the field are obtained. 5th. The cost of manuring with Potash is thus lessened, as the cheaper Potash salts, on account of the magnesia they contain are better for dissolving in this manner. 6th. The expense for plaster which otherwise would have to be employed is obviated." [Dr. Stammer here goes on to prove the value of the sugar beet refuse, and gives the three methods of applying it usually employed, viz: leading the liquids in pipes to the field from a reservoir, carrying it there in barrels, etc., or burning it and then applying the ashes. The first is too expensive for general use, and the the latter causes a loss in nitrogen A. R. L] * * * * * *

"If we ask what quantity of the above recommended manures should be used, surely no farmer would expect a special, universally applicable answer, and I will only recall the *general rule that it is always desirable, if not actually necessary, to restore to a field all the mineral elements of plant food, and from 2 to 3 times the amount of Ammonia removed by the crop.* I will further remark that an excessive application of Potash and Phosphoric Acid (the *cost* of Ammonia will insure that the above given proportion is not exceeded.) has no injurious effect upon the beet, at least not within the limits caused by errors in calculation, or mistakes in practice. On the other hand writers are beginning to agree that excessive application will not increase the yield in the same proportion. * * * * *

* * * * "In speaking of the purely agricultural part of the work of sugar beet culture, I will only point out the importance, the necessity of *deep cultivation*, and though the subsoil, according to its character, need not always be turned up, it *must* be pulverized and drained as well as possible—an *axiom* which cultivation by steam has fixed beyond a doubt." * * * * *

"I think I can not better close this short consideration of the most important points in the development of the beet, than by giving the most important rules in a brief and concise form:

1. Be exceedingly careful in choosing your land and your seed.

2. Spare no pains in applying the manure. For this purpose take into consideration, not only the debtor and credit sides of the *yield* of the field, but also the compensation that the *ground* requires for the constituents removed by the harvest, and their proper return in manure.

3. A rotation of crops must be observed, and such fields kept out of the number used for beets, which show their unsuitableness for beet culture.

4. Beet culture must not be on too large a scale, when one wishes larger harvests and good beets, and larger rather than smaller harvests of grain, than he obtained before going into the beet culture.

5. The preparation of the soil must take place at the proper time, in a proper way, and with proper tools.

6. The seed should be sown as early as the state of the ground and the climate will allow.

7. Be not too tardy in pulling up the beets.

8. The *hoe* should be used as often and as much as possible.

9. The harvest must not be put off. * * *

10. Never cease to observe and learn.

11. Protect the birds, which destroy the hurtful insects, and wage against their enemies a ceaseless warfare."

This excellent advice of Dr. Stammer, embodying the experience of French and German agriculturists, contains much by which we may profit, not only in our experiments with the beet, but also in our general farming.

As before said, the cultivation of the sugar beet has spread to all parts of Europe, and but slight trouble seems to have arisen on account of the difference of climate, for we find the beet growing, and manufactories running in Russia, Sweden, Bohemia, Austria, France, Germany, Holland, Belgium, etc., etc.

In this country nearly every one of the northern States and many of the western have made greater or less experiments on growing the beet, and have usually stopped there. In many instances a fair per centage of sugar was found in the juice, even Canada comparing very favorably with the old world in that respect. Statistics of these experiments are not easily accessible, but I will give below some of the results attained in several States, and will refer the reader to the various State reports for the details.

Average amount of sugar in beets raised in the following localities:

				Per cent. sugar.
Westchester county, N. Y.,	1872,			8.70
Dutchess county,	"	"		10.97
Washington " (a)	"	"		11.70
" " (b)	"	"		9.50
Herkimer "	"	"		11.00
Orleans "	"	"		12.40
Amherst, Massachusetts,	1870,			12.70
" "	1871,			10.79

Amherst, Massachusetts,	1872,	....	7.37
Montreal, Canada,	1973,	....	8.86
Bridgeville, Delaware,	1876,	.. ..	2.75
Camden,	"	"	.... 7.40
Newark,	"	"	.... 3.70
Seaford,	"	"	.... 2.00
Wyoming,	"	"	.... 5.50
Wilmington,	"	"	.... 3.00
Harbeson,	"	"	.... 2.88
Milton,	"	"	.... 3.90
Dover,	"	"	.... 4.40
Felton,	"	"	.... 4.75
Ellendale,	"	"	.... 2.00
Milford,	"	"	.... 14.70
Lincoln,	"	"	.... 3.00
Harrington,	"	"	.... 5.10
Pleasant Hill,	"	"	.... 7.74
Falkland,	"	"	.... 13.00
Farmington,	"	"	.... 5.70
Woodstown, New Jersey,	"	....	4.30
Pennsgrove,	"	"	.... 3.90
Pedricktown,	"	"	.... 4 20
Sharpsburg, Maryland,	"	....	6.20
Omaha, Nebraska,	"	....	13.50
Lincoln, "	"	....	13.50

(13 other localities in Nebraska gave an average of over 15.50 per cent., the highest being 15.61, the lowest 7.20 per cent.

Virginia Agricultural Experiment farm, 1872, (a) 13.72
" " " " " (b) 10.17

There have been many other experiments in growing the beet in different States, but they do not differ in general results from those cited above. It will be seen that in many cases very excellent results were obtained, and yet in spite of this fact there has been comparatively little done in the line of manufacture. I have not thought it necessary to cite the variety of seed used in the several cases, nor the character of soil and cultivation employed.

MANUFACTURE OF BEET SUGAR.

The history of the efforts to make sugar profitably from the beet in this country can be very easily told. The following are the principal experiments in that direction:

David L. Child, of Northampton, Mass., made 1,300 lbs. of sugar from beets grown on his own farm in 1838. Yield, 13 tons of beets per acre, at a cost of $42.

In 1853 Gennert Bros., from Germany, started a beet farm of 2,400 acres, at Chatsworth, Ills. The land "analyzed well," yet failed to yield satisfactory results. Drought, poor seed, floods, &c., also militated against them, and in 1870 they removed to Freeport, Ills., and, if I am not mistaken, have but lately closed their factory, having produced 200,-000 lbs. of sugar in 1870, at a reasonable profit.

In 1867 a company was formed in Wisconsin, at Fond du Lac, under the lead of Messrs. Bonesteel and Otto, but on a small scale, the works having a limited capacity. They have recently consolidated with a California company which is still working successfully in that State.

In 1870 a co-operative company of farmers started a small factory, and were quite successful, at Black Hawk, Wisconsin. A deficiency in their water supply seems to have been their greatest drawback.

The largest and most successful experiment was instituted in California. In 1860 Mr. Speckman made an attempt at beet culture near San Francisco. The soil was not suitable and he abandoned the enterprise. In 1869 Mr. Wentworth instituted another experiment at Alvarado, and succeeded in extracting from his beets several hundred pounds of sugar. Capitalists became interested and a company was formed which, under the management of General Huchison, has had quite a success. The two German experimenters from Fond du Lac, Wisconsin, Messrs. Boncsteel and Otto, were taken into the company, as already stated. Drought and other causes interfered with their success at times, but in 1871 they reported an average yield of 15 tons of beets per acre, and a product of 1,000,000 pounds of sugar. Another company is now formed in California, and the industry seems to have gained more of a foothold on the Pacific coast than elsewhere.

Besides the above experiments many individuals have raised the beet and extracted the sugar on a smaller scale. Prof. Goessman, with apparatus improvised for the occasion, obtained a yield of from 8 to 9 per cent. of sugar from beets grown in Massachusetts, or at the rate of nearly 2,000 pounds of sugar per acre.

The number of failures to make the business pay has been due to a variety of causes; prominent among them a lack of sufficient capital to outlive the unavoidable delays and expense of getting a good start, neglecting to determine beforehand how *cheaply* a good beet can be raised, &c., &c. Bad management, too, had its share of the blame. One company failed, I am told, because they selected for the site of their works the summit of a hill; very picturesque, to be sure, and giving a fine outlook over their acres of growing beets, but unfortunately the extra expense of carting all their fuel and beets up hill and pumping up all their water ate up the profits, and the company failed. Another company bought the works, moved them down the hill and are now said to be doing tolerably well.

There are many sanguine people both in this country and in Europe who point to the sugar beet industry as one of great importance to America, and embodying the potential elements of great wealth to our people. And they point to our climate, soils and improving methods of agriculture; to the comparatively limited area where the cane can thrive, in support of their views. But there are others who point to the price of labor, the expensive processes compared with those necessary for the raising and working up of the sugar cane, and the small profits now being made by our cane sugar manufacturers and refiners, and declare the whole idea a snare and a delusion.

To show the profit and loss side of the question, I will append a calculation made by Mr. II. P. Humphrey, of Philadelphia, a distinguished sugar chemist, who has given much thought and time to the subject. He has addressed a circular to capitalists and others in the hope that a care-ful and thorough experiment, backed by capital, may be made to prove finally the possibility or impossibility of planting this new industry firmly in our midst.

After quoting the statistics I have already recorded to show the amount of beet sugar produced in the world in 1865–'76, Mr. Humphrey goes on to say :

" The following table represents the statistics of the German Empire in regard to the beet sugar industry, as gath-ered from data in " Stammer's Jahresbericht," the most re-liable authority obtainable :

	1 Amount of Beet Centners, (110 lbs.)	2 Amount Sugar Centners.	3 Amount Molasses Centners.	4 Tax per Ctnr. of Beets. (cents.)	5 Amt. received as taxes by Government.	6 Per cent. sugar extracted.	7 No. Ctrs. Beets to produce one Centner Sugar.
1836–'37	506,923	28,162	21,789	.....	....	5.5	18.
1841–'42	5,131,516	284,102	....	.59	....	....	....
1846–'47	5,633,848	402,418	169,015	3.54	200,000	7.14	14.
1856–'57	27,550,208	2,071,579	633,678	14.16	3,912,271	7.52	13.
1866–'67	50,712,709	4,024,818	1,242,461	17.75	10,001,508	7.9	12.6
1873–'74	70,575,000	5,779,442	....	18.93	13,389,827	8.2	12.2
1874–'75	55,072,412	5,011,589	....	18.93	10,525,207	9.1	10.99

Column 1 shows the advance which has been made in the industry since 1836. Columns 4 and 5 show the tax which has been levied and the amounts which have been realized by the Government. The average yield of an acre in Germany is 11.7 tons. The tax paid upon this quantity is $45. This amount would be a great offset to the greater cost of cultivation in this country. No data can be found to establish at what price beets may be raised here. Estimates have been given, which vary all the way from sixty-four cents to four dollars a ton. The average of the results of the experience of eighteen persons is two dollars and forty-two cents per ton. (See "Scientific American," April 3d, 1869.) These estimates, I think, should not be relied on, as the cost would probably reach three dollars per ton; there are also no sufficient data to show the amount of beet roots which can be raised to acre in this country.

The following tables will elucidate these points as regards Germany, France and Russia. These estimates were made some years ago, but will serve to give a general idea of the amount raised per acre:

COST OF PRODUCTION AND THE DIVISION OF EXPENSES INCURRED PER ACRE.

	GERMANY.	FRANCE.	RUSSIA.
Rent and manure...............	$ 18.73 ⎱	$ 38.31	$ 12.39
Cost of production.............	14.28 ⎰	......	12.39
Tax................................	42.30	49.58	13.52
Cost of manufacture........ ..	50.47	69.11	50.70
	$ 125.78	$ 157.00	$ 89.00
Tons of beets per acre.........	11.6	17.9	9.2
Per cent. sugar extracted....	8.	6.	6.
Pounds sugar extracted per acre	2,078	2,403	1.236

By reference to column 6 of preceding table it will be seen that the amount of sugar extracted is nine and one-tenth per cent., which is the amount extracted at the date of this latter table.

I would also call your attention to columns 6 and 7, as showing the gradual increase in the extraction of sugar from the beet, owing to the production of a better quality of beet, as well as the improvements made in the method of manufacture. As regards the possibility of our being able to raise a beet of good quality, there is very little doubt when we consider that excellent sugar beets have been produced here experimentally, and that the beet flourishes in Europe in such a variety of climate, from Italy to Russia and Sweden. In Sweden, where the season is very short, beets having a high per cent. of sugar are produced in paying quantities. I would again refer, whilst considering the probability of the production of the beet at a sufficiently low figure, to the enormous sum paid as tax—from $34 to $54 per acre.

As regards the comparative yield of the beet and cane : In Louisana the average yield per acre is about seventeen tons, and five per cent· is extracted :

	pounds.
Amount of sugar produced per acre in Louisiana,	1,904.
Amount of sugar produced per acre in foreign cane districts,..................................1,600 to 4,800.	
Amount of beet sugar produced per acre in Germany, ...	2,078.
Amount of beet sugar produced per acre in France,	2,403.
" " " " " Russia,	1,236.

Only one-eighth of the amount of sugar consumed in the United States in 1875 was produced in Louisiana.

The following is an estimate which I have made upon the probable returns of a manufactory consuming fifteen million pounds of beets in one hundred days. The manufactory would cost about sixty thousand dollars. Stock of bone black (75,000 lbs) $2,250.

EXPENSES.

16,350 days labor at $1.20		$19,610
Superintendent, ... 3,000		
1 boiler,.................. 1,500		
1 defecator, ... 1,200		
Book-keeper and clerk, 3,000		
Engineer, ... 1,000		
Carpenter, &c., ... 1,000		
Total skill labor..........		$10,700
1,100 tons of coal at.... $3.50		3,850
Taxes, Insurance,........		2,000
Bone black loss...........		750
4,500 bushels of lime at $28		1,260
Int. on working capital		3,500
6,696 tons beets at $4		26,784
Incidentals, ...		10,000
Interest on plant..................		4,200
Interest, total,....................		$28,654

RETURNS.

1,200,000 lbs. 8 per cent., yield at $8.25, ...	$99,000
1,227 tons of press cake at $4 a ton, ...	4,908
Molasses,...	1,825
Residues as fertilizers, ...	1,500
Total, ...	$107,233
Total expenses, ...	82,654
Profit, ...	$24,579

I have considered four dollars per ton a fair price to pay

the farm for the beets. Should it be possible to raise the
beets for three dollars a ton, there would be for a farm of
578 acres the net profit of $6,704 which is to be added to the
above profit if the manufacturer cultivates his own beets.
8¼ cents per pound is a low estimate for raw sugar. Several refiners whom I have consulted consider the value to
be 8¾ cents. I have considered the yield of sugar as 8 per
cent., although last year in Germany in was 9 1-0 per cent.

I have calculated the press cake as worth four dollars per
ton. This is a low estimate, Dr. Goessmann, a gentleman
who is perhaps better acquainted than any one in the country with the facts which bear upon the industry, considers
that the value of the press cake is $17.40 per acre. This
would make the total amount for 578 acres $9,157 instead
of $4,908. I have thought better to choose the latter sum,
as there would be a certain prejudice to be overcome before
the farmers would be willing to pay the real value.

The foregoing are a few of the facts which I have chosen
to present to you, and which have been carefully selected,
and I think in respect to the estimates, fairly stated. My
views as regards the establishment of this industry are as
follows:

Starting out with the idea that the important point to be
established is, how cheaply can beets of good quality be
produced in this country, I would have a company formed
of capitalists who are willing, if the preliminary experiments should prove successful, to furnish about $150,000
capital. Two or three farms of from twenty to twenty-five
acres should be hired, in sections where the climate and
soil appear favorable. These should be cultivated at least
two years with beets according to the methods adopted in
Germany and France. It appears to me of no use to trust
the raising of beets to farmers alone. The experiment has
been tried repeatedly and it has proved almost impossible
to overcome their prejudice as regards the proper method

of cultivation. The amount of beets produced, the per. cent. of saccharine matter, &c., and the total cost should be carefully noted. If it was desired, the raw sugar could be extracted, although this is not essential, as the subject has been so well studied abroad that the amount of sugar obtainable from a beet of a certain quality could be very closely estimated. From these experiments the amount and cost of raising the beet could be fairly calculated; if the results should be satisfactory, the land could be purchased and the manufactory built. The experiment would not be expensive, as the crop of beets could be sold and go far to repay the outlay. It is of course not probable that as satisfactory results as are figured will be obtained for several years, but there seems to be little doubt that Government and State aid might be obtained. Already in Canada they have offered $7,000 a year and exemption from taxes for ten years to the first manufacturer. In Maine the same inducements are offered, one cent a pound until the amount reaches $7,000; with that surety capital would run small risk in making the venture."

I have no doubt at all that such experiments as Mr. Humphrey proposes will be undertaken, and why shall not North Carolina prove so alluring a field that they will be undertaken here?

THE SUGAR BEET IN NORTH CAROLINA.

It was without doubt an easy conclusion to reach that somewhere within the ample borders of the "Old North State" the sugar beet—or indeed almost any other plant— might find itself at home, and prosper, embracing as she does nearly every variety of climate, and infinite diversity of soil, with an average temperature corresponding with that of the best beet growing portions of Europe.

These and other natural advantages, together with the results of a small experiment made in Wake county in 1876, induced the Department of Agriculture to institute the experiment on a broader field, and with this end in view, they procured from a reliable source a quantity of seed of two kinds: the French or "Vilmorin," and the Silesian or "Imperial." These seeds were sent out to 100 prominent farmers, in 34 counties, embracing nearly every variety of soil and climate represented in the State. The following instructions were sent out with the seeds:

"Circular to Experimenters in Sugar Beets.

NORTH CAROLINA, }
DEPARTMENT OF AGRICULTURE, }
RALEIGH, April 20th, 1877. }

Sir :—Encouraged by the results of an experiment, made in Wake county last year with the sugar beet, the Board of Agriculture were induced to buy a lot of imported seeds for purposes of experiment in different sections of our State. It is confidently believed, that should a fair test be made the result would show, that here, as in France and Germany, it would be one of the most profitable industries of our people. Nearly half the sugar consumed by the civilized world is made from the beet and it is thought that the adaptability of our soil and climate to its successful culture, would at no distant day, place our State on the list as among the largest producers of this great commercial commodity.

With the view of introducing this important industry, you have been selected as a proper person to test it in your county. The following suggestions we ask should be *observed rigidly*, as they are founded on the experience of 50 years in Europe, and are indispensable to success.

Soils.—Newly cleared, heavy clay, wet or salt lands are

unsuited to the beet—any good wheat lands, light, rich and loamy; or in other words, any place that would make a good garden spot would suit the beet.

Preparation and Manure.—Plough or spade at least 15 inches deep, and pulverize thoroughly, putting on broadcast any commercial fertilizer known to be good for vegetables, at the rate of 400 pounds per acre, or ashes at the rate of 25 bushels per acre. *Be sure not to use stable or barnyard manure.* The object of deep cultivation is to cause the beet-root to grow entirely below the surface, the part above being injurious to sugar making, and if the root should grow above the surface it must be kept covered with earth.

Seed-planting.—Soak in water 24 hours, and as soon as you see signs of sprouting roll them in wood ashes or plaster, and plant not more than one inch deep, and thick enough in the drill to leave the plants from 8 to 12 inches apart after thinning. Have the rows from 18 to 22 inches apart. Remember that *large* beets are poor in sugar, and it is the *percentage of sugar we wish to determine.*

Cultivation.—Should be deep and thorough, and should begin as early as practicable, keeping the ground loose and clear of weeds—thinning out or transplanting as may be required to secure a proper stand.

Maturity of Beet.—This will be ordinarilly about five months after planting. The proper time for gathering may be ascertained by the leaves turning yellow or looking flabby—or perhaps better still by cutting a root in slices with an iron knife, and if the surface cut does not change its color, or if any, but little, it is time to take them up. If, however, the surface should turn first red, then brown, and finally quite dark it is too soon. In harvesting, particular care should be taken not to cut or bruise them, and they would do better, if the weather be favorable, to lay them in piles on the ground, and cover with the tops to protect them from the sun, for three or four days.

Preservation — In our climate the usual methods adopted for keeping the ordinary beet or potatoes will answer.

Report to be Made.—You will please keep correct notes of your process of treatment from the time you begin the preparation of the ground— kind of soil and subsoil—kind and quality of manure used—mode of cultivation—estimate of the number of bushels per acre, &c.

It is exceedingly important that this report be correct.

Packing the Lot for Analysis.—As soon as you gather them, you will select carefully not less than two bushels, taking particular care to select such as have the *roots* and *tops entire* and *unbruised*, and that are of *average size* and *well matured.* Do not wash them, but rid them of the dirt as best you can without breaking the roots, and pack them in a good strong crate or box, so made as to admit passage of air. Mark the box plainly, " *Department of Agriculture,*" Raleigh, N. C., and send it by Express. We will pay all charges. In packing use green leaves or grass. This lot is designed for analysis by the Agricultural Chemist, and whatever expense is incurred in packing and shipping will be paid by this Department.

L. L. POLK, *Commissioner.*"

As already stated, we are disappointed in the meagre returns received and the general low per centage of sugar obtained. As will be seen from the letters of some of the experimenters, there are several general reasons for these results; unfavorable season, the ravages of insects, &c., &c. While the unavoidable but unfortunate delay in sending out the seeds is another cause of the comparative failure.

I will now give the results of my analyses of the samples received.

ANALYSES.

In each case I have determined :

(1.) The weight of the beets.

(2.) Specific gravity of the juice.

(3.) Water in the juice.

(4.) Cane sugar in the juice.

(5.) Substances in juice other than cane sugar (by difference.)

The beets were carefully cleansed by brushing and rubbing, without washing, and grated by hand on a large tin grater. The pulp was subjected to pressure in thick cloths, or in an iron screw press.

The method of analysis was, briefly, as follows: The water was determined in the usual way by heating a certain portion mixed with a weighted amount of pure, dry sand, at 212° F. until the weight remained constant.

The sugar was determined in an accurately measured portion of the juice by means of an excellent "polariscope."*

The determinations were duplicated in almost every case, and the figures given are the average of all observations.

LOT No. 1.

Raised by Mr. J. C. Pass, Faison's, Duplin county. Received in September.

Weight of largest........................... 15½ ozs.
  "   " smallest .... 6½ "
Average of twenty beets.. .... 9.6 "

---

*I am deeply indebted to Dr. Arno Behr, of the sugar refinery of Messrs. Matthiessen & Wiechers, Jersey City, for the loan of a valuable instrument, by which I was enabled to commence work promptly without annoyance from a delay in receiving the instrument ordered by the Department from Dr. Scheibler, of Berlin.

Specific gravity of juice.................................... 1.0417

PER CENT.

Water............................... .... 86.63
Sugar 6.46
Solids other than sugar 6.91
_____
100.00

A "check" analysis on a second lot of the same beets gave
the following results:

Specific gravity...................................... 1.0396
Sugar . ... 6.44 per ct.

Mr. J. C. Pass reports upon this lot raised by him as follows :

"It is possible that the soil on which the beets were
planted, though well drained and very fertile, contains an
excess of saline and alkaline substances. The specimen sent
is *not* above the average.

Land planted, 40x89 feet, (1-12 acre.)

Fertilizers used, acid phosphate (Navassa), 40 lbs.; wood
ashes, about one bushel.

Quality of land, rich loam.

Mode of preparation, flushed with a one-horse plow and
sub-soiled 15 inches deep, and thrown into ridges 22 inches
apart.

Quantity of seed sown, quarter of a pound, *less* one table-
spoonful. Seeds put into water to soak the 25th April,
rolled in wood ashes and planted the 26th and lightly covered, from which a stand of about 85 per cent. was obtained
(7 inch drill space desired). The re-set beets from thinning
the hills were of no value.

Cultivation consisted in plowing *one* time and hoe-worked
four times.

The cut-worm attacked the plants in spring. Some of the tubers commenced decaying in the latter part of July.

The bugs attacked the tops about the middle of August, and have proved about as destructive to beet tops as the army worm to the cotton plant.

Dug 910 lbs. of roots, including debris of tops left by the bugs."

LOT No 2.

Raised by Mr. J. W. Pelletier, Morehead city, Beaufort county. Received in September.

Total weight............................	7 lbs.
Weight of largest..........................	1 lb. 2¼ ozs.
" " smallest..........................	5½ ozs.
Average of 10	11 2 "
Specific gravity of juice...................	1.0390

	PER CENT.
Water....................................	87.28
Sugar	5.12
Solids other than sugar....................	7.60
	100.00

Mr. Pelletier reports as follows:

"Received seed, May 8th, 1877.

Soil, gray sand, with yellow sub-soil

Preparation, broke up ten inches deep.

Manured with ashes at the rate of 20 bushels per acre, with same amount of cotton seed. Planted seed 11th May. in drills two feet apart and ten inches in drill.

Cultivation.—Plowed 25th May, 16th June, 9th July and 28th July, with a small turn plow, and followed each plowing with hoe.

Yield, 150 bushels per acre.

I have no experience in raising the sugar beet, but the ordinary beet, planted the first of March, would yield at

least twice as large a crop as when planted the 11th of May.
The land should have been plowed at least a month before
planting."

Mr. J. H. Swindell writes as follows:

"I turned over my seed to Mr. John W. Pelletier, one of
the best farmers of our section. I would add to Mr. Pelletier's report that we have had this season entirely too much
rain for beets to do well. I am satisfied he would have done
much better but for this and not having received the seed
earlier."

LOT No. 3.

Raised by James Norwood, Esq. Hillsboro,' Orange county.
Received in September.

Total weight.......................................	25¼ lbs.
Weight of largest...............................	3 "
" " smallest..............................	1 "
Average weight of 13...........................	2 "
Specific gravity of juice......................	1.0322

	PER CENT.
Water	88.54
Sugar..	10.24
Solids other than sugar.......................	1.22
	100.00

Mr. Norwood reports as follows:

" Planted on the 27th April, the seed already sprouted, on
a piece of ground 5 by 33 yards, in rows 2½ feet apart;
when about an inch high gave a dressing of 2 bushels of
half-slacked ashes along the line of plants, and thinned
them to 12 inches apart; then with a good cultivator stirred
the ashes well in. About a month afterwards, sowed 2 bushels
of ashes broadcast and gave the patch a deep and thorough
stirring with a cultivator. Afterwards kept the weeds and
grass out.

The piece of land is not rich, would bring 4 barrels of corn to the acre, is red clay soil; no sand in it; was twice ploughed in the last of the winter, and thoroughly broken up 9 inches deep.

I will deliver a sample of the beets, and you will find them too large, probably:"

LOT No 4,

Raised by Capt. Jno. Hutchins, 4 miles from Chapel Hill, Orange county.

Total weight.. 14½ lbs.
Weight of largest.. 4 "
   "   " smallest.. 1½ "
Average weight of 6.. 2¾ "
Specific gravity of juice........................... 1.0408

	PER CENT.
Water	90.08
Sugar	4.55
Solids other than sugar	5.37

100.00

Capt. Hutchins reports that he did not attend to the planting or cultivation of the beets in person—in fact did not know they were on his plantation until ready to pull. They were planted in rich bottom land, and had little or no care.

LOT No. 5,

Raised by Mr. H. W. Ledbetter, Wadesboro, Anson county.

Total weight.. 18¼ lbs.
Weight of largest.. 3 "
   "   " smallest.. 1 "
Average of 11.. 2 "
Specific gravity of juice........................... 1.0248

	PER CENT.
Water	89.79
Sugar	4.30
Solids other than sugar	5.91
	100.00

" I have taken up, placed in a box, and will ship in a few days, as directed, ½ bushel of the sugar beets. As you know I did not get the seed until May—they should have been planted several weeks sooner—I planted six short rows in my sweet potato patch, a light loam, rather sandy. Planted 11th of May, used Whann's Raw Bone Superphosphate in drill, about 300 lbs per acre, gave same cultivation as I did cotton adjoining. There came up about half stand, and grew finely ; had no rain until 3rd of June. New beets came up, the older ones were then about 6 inches high ; the younger ones never did much, the sun was too hot for them. The beets seemed to do well until the dry hot weather of August, when the tops seemed to die and fall off. They are putting out again new, and seem to be taking the second growth. I estimated the yield to be about 210 bushels per acre. If planted earlier, with suitable preparation of good manure, and good cultivation, the yield would have been 3 times as large. The land has a clay sub-soil, red, at about 12 or 14 inches from top. Sorry I did not get the seed in time to make a more complete experiment ; will try again next year."

LOT NO. 6.

Raised by Dr. G. W. Blacknall, Raleigh, Wake county. Received in September :

Total weight	51¾ lbs.
Weight of largest	5¼ "
" smallest	3¼ "
Average weight of 12	4 lbs. 4 ozs.
Specific gravity of juice	1.0182

	PER CENT.
Water	92.85
Sugar	4.55
Solids other than sugar	2.60
	100.00

No report received.

LOT NO. 7.

Raised by Mr. John M. Crenshaw, Forestville, Wake county. Received in October.

Total weight	26 lbs.
Weight of largest	3 "
" smallest	¾ "
Average of 22	2 "
Specific gravity of juice	1.0183

	PER CENT.
Water	89.06
Sugar	6.97
Solids other than sugar	3.97
	100.00

Mr. Crenshaw reports:

"They were received late in the season, without any idea of their mission, and were planted and not much attention given them."

LOT NO. 8.

Raised by —— ———, Tarboro, Edgecombe county. Received in October.

Total weight	23¾ lbs.
Weight of largest	2 "
" smallest	¾ "
Average weight of 25	1 lb. 6 ozs.
Specific gravity of juice	1.0498

	PER CENT.
Water	87.99
Sugar	6.30
Solids other than sugar	5.71
	100.00

No report.

LOT NO. 9.

Raised by A. M. McPheeters, Raleigh, Wake county Received in October.

Total weight	18½ lbs.
Weight of largest	1 lb.
" smallest	6 ozs.
Average weight of 34	11 "
Specific gravity of juice	1.0378

	PER CENT.
Water	86.24
Sugar	10.97
Solids other than sugar	2.79
	100.00

No report·

LOT NO. 10.

Raised by W. M. Blackwell, Oxford, Granville county. Received in October.

Total weight	19½ lbs.
Weight of heaviest	1¼ lbs.
" lightest	7 ozs.
Average weight of 32	14 ozs.
Specific gravity of juice	1.04270

	PER CENT.
Water	84.96
Sugar	11.37
Solids other than sugar	3.67
	100.00

A check analysis on another lot gave:

Specific gravity.................................... 1.04277

Sugar .... 11.46 pr. ct.

Mr Blackwell reports:

"I plowed the land about six inches deep with a cast turning plow, and followed in the same furrow with a coulter. breaking the land about 14 or 15 inches deep, and applied 700 lbs. to the acre of "Dixon's Compound," composed of equal parts of Peruvian guano, dissolved bone, plaster and salt. I planted the seed the 4th of May. As soon as necessary, I thinned them to 8 or 10 inches apart in the drill, rows 22 inches apart. The first working was done with a harrow, the second working a small turning plow was run, throwing one furrow to the beets, and harrow in middle of the row, which was all the plowing I did to them; each time the hoes followed the plow, chopping out all grass and weeds, and leaving the land nearly level. The seed was planted on sandy soil with yellow clay (subsoil?) The yield was only 140 bushels to the acre. We had the worst seasons I ever saw, owing to excessive wet. My crops of all kinds were seriously injured, as we had more rain in my immediate neighborhood than any other part of the country· I gave one of my neighbors, W. B. Crews, some of the seed, and his land was better adapted to beets than mine. he using ashes as a fertilizer, his yield was 250 bushels to the acre."

LOT NO. 11.

Raised by Mr. J. W. Wilson, Morganton, Burke county Received in October.

Total weight......................................	30¼ lbs.
Weight of largest	3 "
" smallest................................	¾ "
Average weight of 24.............................	1¼ "
Specific gravity of juice........................	1.0175

	PER CENT.
Water	92.14
Sugar	5.91
Solids other than sugar.........................	1.95
	100.00

No Report.

LOT NO. 12.

Raised by ————————, Chapel Hill, Orange county. Received in November.

Total weight...	49 lbs.
Weight of largest	2¾ "
" smallest..............................	12 ozs.
Average weight of 36	1 lb. 12 ozs.
Specific gravity of juice	1.0247

	PER CENT.
Water	94.07
Sugar...	3.35
Solids other than sugar	2.58
	100.00

These beets were grown upon very rich soil, (in a garden spot) and were fertilized with ashes.

LOT NO. 13.

Raised by J. W. Wissler, Lockville, Chatham county. Received in November, marked "Imperial, No. 4."

Total weight, (but 2 in the lot) ..:............	4 lbs.
Average weight	2 lbs.
Specific gravity of juice	1.0396

	PER CENT.
Water	90.99
Sugar	5.51
Solids other than sugar	3.50

100.00

Mr. Wissler sends samples of 6 lots, and his report accompanying them will be given after the analysis of all the samples.

LOT NO. 14.

Raised by Mr. J. W. Wissler, Lockville, Chatham, county. Received in November, and marked " French, No. 2,' only *two* beets in the lot.

Average weight	1 lb. 6 ozs.
Specific gravity of juice	1.0501
Check " " "	1.0502

	PER CENT.
Water	85.47
Sugar	10.82
Solids other than sugar	3.71

100.00

LOT NO. 15.

Raised by Mr. J. W. Wissler, at Endor Furnace, (near Egypt) Chatham county. Received in November, and marked "Imperial No. 6," only *two* in lot.

Average weight	1 lb. 9½ ozs.
Specific gravity of juice	1.0380

	PER CENT.
Water	91.33
Sugar	5 27
Solids other than sugar	3.40

100.00

LOT NO. 16.

Raised by Mr. J. W. Wissler, at Lockville, Chatham county. Received in November, and marked " Imperial No. 1," *two* in lot.

Average weight	1 lb. 10½ ozs.
Specific gravity of juice	1.0307
	PER CENT.
Water	93.21
Sugar	5.27
Solids other than sugar	1.52
	100.00

LOT NO. 17.

Raised by Mr. J. W. Wissler, at Lockville, Chatham county. Received in November, and marked " French No. 5," *two* in lot.

Average weight	1½ lbs.
Specific gravity of juice	1.0522
	PER CENT.
Water	87.61
Sugar	11.22
Solids other than sugar	1.17
	100.00

LOT No. 13,

Raised by Mr. Wissler, at Lockville, Chatham county. Received in November. and marked "French No. 3." *Two* in lot.

Average weight	1 lb. 5 ozs.
Specific gravity of juice	1.0436

	PER CENT,
Water	90.08
Sugar	7.14
Solids other than sugar	2.78
	100.00

Mr. Wissler reports as follows:

"No. 1, (Lot No. 16), is the Imperial seed, put in soak on the 11th of May and planted on the 12th, in a lot on river (low ground). This piece of ground had been used for several years as a cucumber patch. Top-dressed heavily with "Phuine." Plowed deep (about 10 inches) first time, on the 11th, and again on the 12th, before planting. The ground was nicely pulverized and in good condition for planting. After planting on the 12th (May) we had no rain until the 9th of June, when they came up nicely, and I had them replanted, when it again got dry, and I suppose we had no rain for at least six weeks. Owing to this, I failed in having a stand anywhere.

I worked the beets about the same as I would a crop of corn. After the season got better they commenced growing and were still growing finely when I had them pulled up on the 23d of October. On that night we had a heavy frost, and I was fearful it might injure them to leave them stand longer.

No. 2, (Lot No. 14), is the French Beet, planted on same ground, and same as No. 1.

No. 3, (Lot No. 18), is the French Beet, planted on same day and treated similarly as Nos. 1 and 2, except that it was planted on a stiff clay. This lot had been in clover for two years, heavily top-dressed with stable manure, and broke up about the 15th day of February, and again top-dressed with "Phuine" and plowed on the 11th of May. Just before planting, this ground—owing to the dry weather—got so hard that I never, even by transplanting, got a stand; but late in the summer and fall they seemed to grow well.

No. 4, (Lot No. 13), is the Imperial Beet, planted on same ground as No. 3, and treated similarly; all these beets were gathered on the 23d of October.

No. 5 is the French Beet, planted at Endor Furnace, (near Egypt), on the 16th of May, on a nice, mellow soil, neither sandy nor stiff. The ground had been plowed about six weeks before, and I had it plowed again just before planting. After drawing the furrows (which I did in all cases with a 'scooter' plow), I had it well sprinkled with ashes, this being the only fertilizer used in Nos. 5 and

The seed had been soaked forty-eight hours, and was nicely sprouted when planted. If any difference, the season was drier at Endor than at Lockville, and yet I must consider my success the best there, the beets growing smooth and of a uniform size of form, 2 to 3 lbs. each.

No. 6 is the Imperial Beet, treated similar to No. 5.

I had still made another experiment at Buckhorn, but must confess to a failure there. The beet, through neglect and dry weather grew very slowly, and was not over half pound beets.in October, when gathered, and so green that they soon dried and shriveled up.

The last of March or beginning of April is the proper time to plant here, when the beet will have root enough to stand the drought and hot weather of July.

One difficulty I had was to keep the beet covered, it seemingly having a tendency to grow out of the ground; for this reason I shall in future plant three feet apart, instead of two. as this year."

LOT No. 19,

This was a separate lot of 7 beets which came in the box with *Mr. Wissler's* samples. Whether a separate lot, or merely specimens from the other 6 lots, I could not ascertain, and so analyzed them.

Average weight.............. .... 1 lb. 13 ozs.
Specific gravity of juice. 1.0421

PER CENT.

Water.................... .... 90.34
Sugar.. 5.96
Solids other than *sugar*......................... 3.70

_____

100.00

LOT No. 20,

Raised by Mr. Columbus Mills, Concord, Cabarrus county.

Total weight.. 27½ lbs.
Weight of largest............ .... 5 "
    "   " smallest.................................... 2¾ "
Average weight of 7............................... 3½ "
Specific gravity of juice............................. 1.0414

PER CENT.

Water... 91.23
Sugar 5.50
Solids other than sugar............................ 3.27

_____

100.00

Mr. Mills reports:

" Planted the last of April, and came up sparsely until the June rains; rows 1½ feet apart; ground, a rich loam. They were worked according to the French mode of cultivation."

LOT No. 21,

Raised by ————, near Chapel Hill, Orange county. Received December 1st.

Total weight.. 16 lbs.
Weight of largest.................................... 2½ "
    "   " smallest...... .... 8 ozs.
Average of 12.. 1½ lbs.
Specific gravity of juice....... .... 1.0390 (?)

	PER CENT.
Water	89.37
Sugar	7.61
Solids other than sugar	3.02
	100.00

No report.

CONCLUSION.

While some of the lots analyzed show a very low per centage of sugar, there are, on the other hand, five (more than one-fifth of the whole over ten per cent., viz:

SAMPLE FROM	PER CT. SUGAR.
Oxford	11.46
Egypt	11.22
Raleigh	10.97
Lockville	10.82
Hillsboro	10.24

Of the remaining sixteen lots more than three-fourths go over five per cent., by no means a very bad showing.

It had been my intention to endeavor, by a careful examination of the results obtained in different localities, and on different soils, and with different management, to be able to point out the cause of failure of one, or the success of another in obtaining beets with a high per centage of sugar. But with the few results that we have been able to gather and the few reports made by the experimenters, we would reach an idle, or at least highly conjectural conclusion. Still, I would recommend our farmers to notice carefully the reports of the experimenters, and compare the results of their labor, and they will oftentimes find valuable hints and suggestions for their guidance next year.

4

I will close with the reiteration of the wish that the result of this report may be to keep alive the interest in our experiments, and that another year may find our people ready to give intelligent and efficient aid to the Department of Agriculture in its efforts to implant this new industry in our State.